Counting

by Maryellen Gregoire

Nancy E. Harris, M.Ed—Reading
National Reading Consultant

capstone
classroom

Heinemann Raintree • Red Brick Learning
division of Capstone

One blue crayon.

One red crayon.

One green crayon.

One yellow crayon.

One purple crayon.

One orange crayon.

One black crayon.